Manufactured by *BMM* in early autumn 1944 *Jagdpanzer 38(t) Hetzer* captured by the US Army in Czech Republic, spring 1945. The curiosity is that the vehicle has the late version of *Kugellafette V* armor plate, road wheels with narrower rubber rims but the early model of the muffler (the second type without additional pierced shelter) and a drive sprocket. Spare tracks installed on the rear plate, the late model of the rear spotlight installed on the left fender and factory-applied national symbol – the white contoured cross with short arms are clearly visible too. The camouflage represents the second scheme superimposed in the production halls of the Prague factory, so called *Hitnerhalt-Tarnung*, consisted of dark green (RAL 6003) and chocolate brown (RAL 8017) surfaces over the sand yellow (RAL 7028) background, replenished with smaller spots causing the light-reflexes. [Icks collection, Patton Museum Library]

Variants

In December of 1943 an idea to design an upgraded version of the *Jagdpanzer 38* armed with *7,5 cm Pak 39 L/48* gun mounted in a special yoke without the recuperator emerged. In the intention of the designers this solution should increase the free space inside the vehicle and allow resigning from cutting out a big opening accommodating this assembly in the front armor plate and the force of recoil was to be absorbed by the vehicle itself. So modified *Hetzer*, designated *Jagdpanzer 38 starr* underwent a series of firing tests at the latest in July of 1944, repeated later at least once (in the first week of September). The results are sadly not known, however it can be concluded that they ended with success. It can be proved even by the fact that immediately after the completion of the tests an order for further two prototypes was placed and on the turn of 1944/1945 the BMM factory completed further ten *Jagdpanzers 38 starr* as so-called zero production batch. The last one, fourteenth, allegedly rolled out of the factory in April 1945.

According to most frequently given information one of the "starrs" of the "0-serie" was sent to Berka training center, but in April of 1945 due to impending Allied troops and the hazard of the capture of the vehicle it was destroyed by a personal order of Adolf Hitler. Of the nine destroyers remaining at Milovitz proving ground eight were converted back to normal *Hetzers*. The fate of the last one, as well as three prototype *Jagdpanzers 38 starr* is unknown. This version of events at least partially conflicts with preserved photographic material which shows that at least five Hetzers of the "starr" variant were used in combat by the Germans on the afternoon of 5 May 1945 in Podbaba

against Czech insurgents. Noteworthy is also the fact that at least one "starr" was on the strength of the unit called *Kampfgruppe Milowitz*, hastily formed on Milovitz proving ground. The personnel of the unit consisted mainly of the instructors and tank soldiers of the *507. s.Pz.Abt.* which after the loss of all *Tiger II* tanks lost any combat capability.

In May of 1944 at *Böhmisch-Mährische Maschinenfabrik* the *Bergepanzer 38* recovery vehicle based on modified chassis of the *Jagdpanzer 38* was designed for *Panzerjäger-Abteilungen* equipped with *Hetzers*. The vehicle was stripped of the arma-

Jagdpanzer 38(t) Hetzer completed in July 1944. The vehicle had the first type of idler wheels and the *Kugellafette V* gun mantlet. Wider rubber rims of the road wheels were characteristic for the specimens manufactured before August 1944. There is also a single part of side skirts, which protected the upper part of tracks. The camouflage scheme, consisted of irregular, blur-edged, dark green (RAL 6003) and chocolate brown (RAL 8017) spots applied on the sand yellow base (RAL 7028). Germany, May 1945. [Kagero Archive]

An experimental version of *Jagdpanzer 38(t) Hetzer* captured by the US Army on the *Rheinmetall-Borsig* proving grounds in Hillersleben, spring 1945. The vehicle conjoins the early production series elements: the first type of muffler with the pierced extra-protection plate, idler wheels as well as the double driver's visor and the road wheels with narrower rubber rims installed from August 1944. The single-chamber muzzle brake at the end of the barrel and the *Kugellafette IV* gun mantlet attract attention too. The vehicle's camouflage is sand yellow (RAL 7028) overall, except of the barrel which was probably in natural steel finish. [Kagero Archive]

ment but was retrofitted with special equipment for evacuation duties. The trials verifying the usefulness of the vehicle for towing a *Hetzer* were conducted in July of 1944 at Kummersdorf. The *Bergerpanzer 38* with chassis number 321072 which was put to the test failed to fulfill the task and after having traveled 39 kilometers with a *Hetzer* destroyer weighing only 14.6 tonnes on tow during 12.25 hours had a gearbox failure. It was not the only worry for the Germans at that time – the fuel consumption of 7.5 l/km was absolutely unacceptable. After a while other *Bergepanzer 38* (chassis number 321073) set out for rescue and finally did a bit better than the predecessor. Before the engine

broke down, it managed to travel 184 km with average speed of 7.2 km/h and fuel consumption of 4.48 l/km. From May of 1944 till April of 1945 a total of 181 *Bergerpanzer 38s* rolled out of the *BMM* factory. In late 1944 the chassis of at least one of them served as a basis for development of an infantry self – propelled gun designated *15 cm s.I.G.33/2 (Sf)*.

In October of 1944 studies on retrofitting the *Jadpanzer 38* with *7,5 cm Pak 42 L/70* long-barreled gun began at *Alkett* company in Berlin. In so rearmed vehicle the use of a new powerplant (*Tatra 103 V-12* Diesel engine rated at 207 hp at 2250 RPM) new gearbox (*ZF AK 5-80*), new turning mechanism and strength-

Jagdpanzer 38(t) Hetzer manufactured by *BMM* and attached to *SS-Panzerjäger-Abteilung 8* of the 8th SS Cavalry Division *Florian Geyer*. The *Kugellafette IV* gun mantlet and the early type of the driver's vision are clearly visible. The gun barrel is protected by the canvas cover in its full length. The vehicle sports the monotone sand yellow (RAL 7028) camouflage scheme. It was the unit commanded by *SS-Brigadeführer* Joachim Rumohr, which became first part of new light tank destroyers in the whole *Waffen-SS*. Hungary, summer 1944. [Bundesarchiv]

Mariusz Motyka • Hubert Michalski • Łukasz Gładysiak • Stefan Dramiński

Panzerjäger 38(t)
Hetzer
& G-13

vol. II

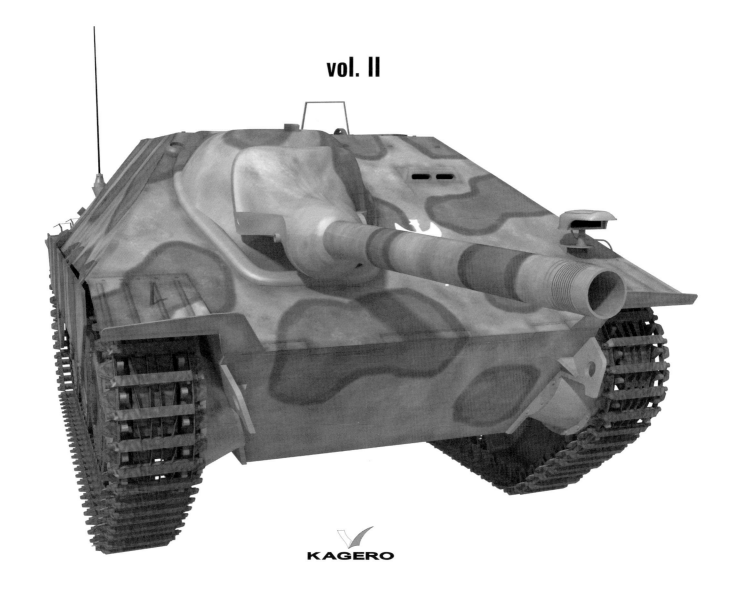

KAGERO

Panzerjäger 38(t) Hetzer & G-13. Vol. II • **Mariusz Motyka** • **Hubert Michalski** • **Łukasz Gładysiak** • **Stefan Dramiński**
• **First edition** • **LUBLIN 2015**

Editors: **Łukasz Gładysiak** • Translation: **Jarosław Dobrzyński** • Color profiles: **Mariusz Motyka, Stefan Dramiński**
• Scale drawings: **Mariusz Motyka** • Design: **KAGERO STUDIO, Marcin Wachowicz**

Oficyna Wydawnicza KAGERO
Akacjowa 100, Turka, os. Borek, 20-258 Lublin 62, Poland, phone/fax: (+48) 81 501 21 05
www.kagero.pl • e-mail: kagero@kagero.pl, marketing@kagero.pl
w w w . k a g e r o . p l

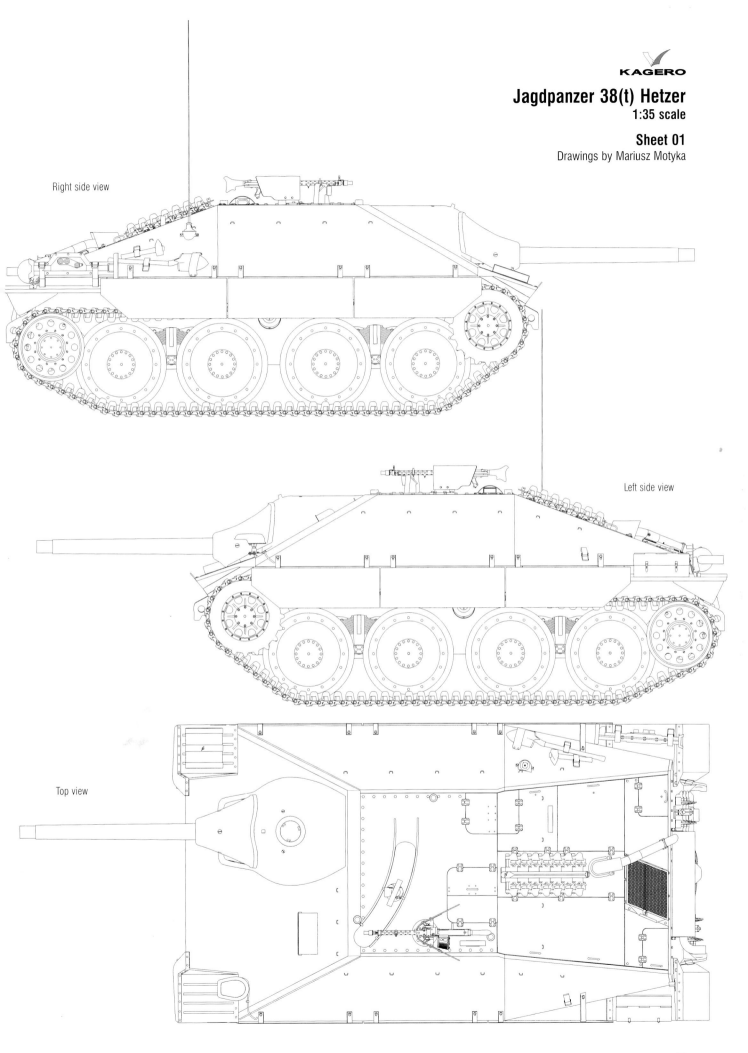

Right side view

Left side view

Top view

Jagdpanzer 38(t) Hetzer completed in autumn 1944, belonged to extemporaneously created *Panzergruppe Beyer*. The late version of *Kugellafette V* gun mantlet and severed by the interior explosion part of the superstructure are clearly visible. There is also lack of the right fender. There are remnants of the white-wash camouflage paint on the armor plates and road wheels. Kolberg, march 1944. [Museum of the Polish Arms Archive]

ening of the suspension was planned. The *Hetzer* designed to this specification was initially named *Jagdpanzer 38 Ausführung "Reich"*, later changed to *Jagdpanzer 38 „d"* and shortly after to *Jagdpanzer 38 D* (some literature claims that the *Jagdpanzer 38 „d"* and *Jagdpanzer 38 D* were two different vehicles). Looking at the design from tactical, technical and logistic point of view one can have an impression that if it was implemented, it would not

do well in combat. If the German engineers managed to fit a gun heavier than the *7,5 cm Pak 39* to already overloaded vehicle, almost for sure the traction characteristics would deteriorate and during downhill drive there would be even a hazard of stability loss. The use of a Diesel engine was not sound either because the fuel oil was hardly available in late 1944. In consequence there could be a paradoxical and pretty imaginable situation that

Jagdpanzer 38(t) Hetzer manufactured by *Škoda* in the second half of 1944. The late version of *Kugellafette V* is clearly visible. The vehicle camouflage consists of wide, irregular, dark green spots (RAL 6003) with red brown (RAL 8012) edges over the sand yellow (RAL 7028) base. Western front, autumn 1944. [Kagero Archive]

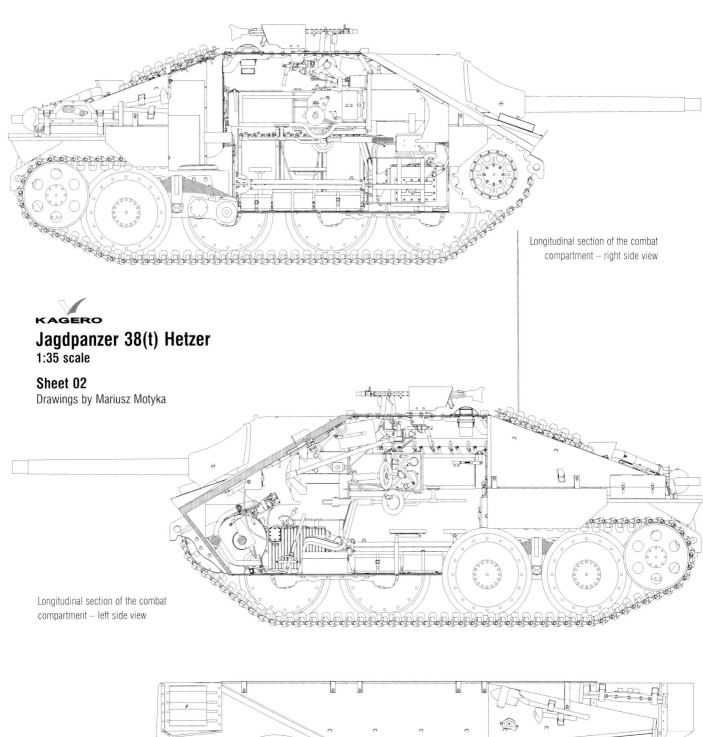

KAGERO

Jagdpanzer 38(t) Hetzer
1:35 scale

Sheet 02
Drawings by Mariusz Motyka

Longitudinal section of the combat
compartment – right side view

Longitudinal section of the combat
compartment – left side view

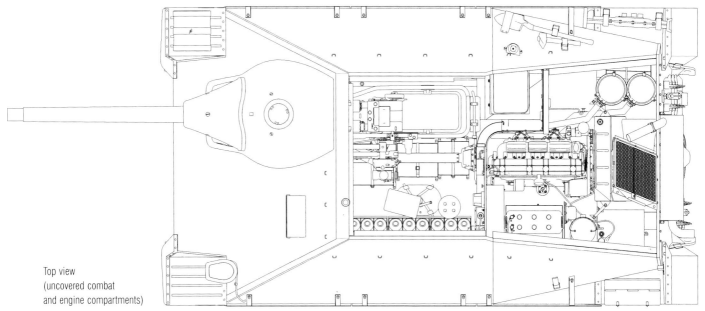

Top view
(uncovered combat
and engine compartments)

Jagdpanzer 38(t) Hetzer **destroyed by the American** *M10* **tank destroyer in the area of Halloville in France, November 1944. The vehicle's camouflage represents the factory-applied** *Hinterhalt-Tarnung* **(so called** *Ambush***) spotted variant. [Kagero Archive]**

already built vehicles would not be used in combat due to lack of fuel. During the studies on the vehicle it was also noticed that the change of the powerplant to a Diesel engine would require designing the drive mechanism from scratch because only the road wheels and tracks could have been adapted from the original vehicle. Taking these circumstances into consideration there is no wonder that the program of the development of the *Jagdpanzer 38 D* ultimately ended with failure, although despite difficulties the Germans took the trouble to build test vehicles. According to most frequently cited information, the falsification of which is currently impossible, in March of 1945 two prototypes were allegedly complete in ca. 80%.

Jagdpanzer 38(t) Hetzer **manufactured by** *BMM* **in the first half of 1944. The vehicle has the early type of road wheels with wider rubber rims and the** *Kugellafette IV* **gun mantlet. Two of the three skirts installed alongside the superstructure edge are clearly visible. The camouflage pattern is a sand yellow (RAL 7028) overall, typical for the early productiona series and has no tactical signs. The western front, autumn 1944. [Kagero Archive]**

Cross-section of the
combat compartment

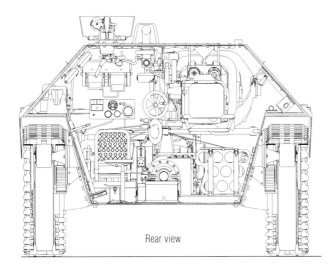

Rear view

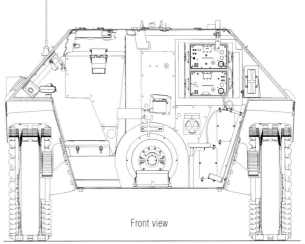

Front view

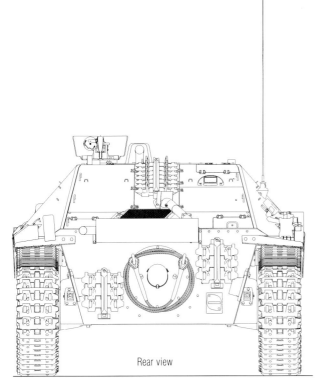

Rear view

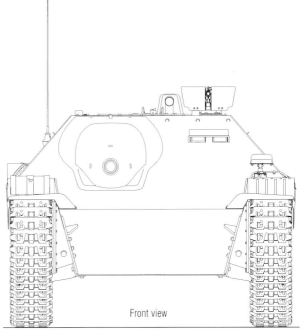

Front view

In early December of 1944 twenty *Jagdpanzer 38s* from the BMM factory were chosen for conversion to self-propelled flamethrowers, called *Flammpanzer 38*. The conversion consisted in the replacement of the gun with the *Koebe Flammenwerfer* flamethrower and fitting of 700-liter incendiary liquid tank, allowing for 60-70 "shots". The vehicle's weight was 13.5 tonnes, the ground pressure was 0,78 kg/cm² and the power/weight ratio was 11.8 hp/tonne. The *Flammpanzer 38s* equipped two self-propelled flamethrower companies (*Flamm-Panzer-Kompanien*) numbered 352 and 353, each of which received

ten vehicles. Both units took part in the Operation Nordwind. The *Flamm-Panzer-Kompanie 352* of the *25. Panzer-Grenadier-Division* went into action on 9 January 1945 during the assault on Hatten. Interesting is that it incorporated the remains of the *Flamm-Panzer-Kompanie 353* in the strength of three *Flammpananzer 38s* since that unit had suffered severe losses in earlier battles (all officers had been killed). Later the *Flamm-Panzer-Kompanie 352* took part in the battle of Rittershofen, where it lost three *Flammpananzer 38s*. On 15 March 1945 the company reported 11 self-propelled flamethrowers in the

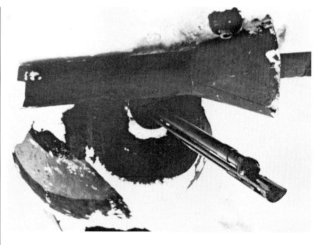

Flammpanzer 38(t) Hetzer **captured by the US Army during the Battle of the Bulge, end of December 1944. The exterior part of the gun barrel connected with the mantlet was removed and the exact** *14 mm Flammenwerfer 41* **flame thrower tube is visible. There is also the second type of the driver's visor installed. The vehicle had been produced by Škoda, which suggests the characteristic camouflage scheme – vertical dark green (RAL 6003) and chocolate brown (RAL 8017) over the sand yellow (RAL 7028) background, but likewise other twenty specimens, armed by** *BMM*. **[Icks collection, Patton Museum Library]**

inventory, including eight roadworthy.

In January of 1945, during the work on the *Jagdpanzer 38 D* it was decided to use its chassis after minor modifications for the development of a whole range of vehicles of various purposes, namely: a reconnaissance vehicle armed with a 2-cm gun in a rotating turret (*Aufklärer mit 2 cm im Drehturm*), a reconnaissance vehicle armed with a 7.5-cm gun with barrel length of 48 calibers (*Aufklärer mit 7.5 cm*), an armored fighting vehicle armed with a 120-mm mortar (*Panzerfahrzeug mit Granatwerfer*), armored personnel carrier capable of carrying a half of an infantry platoon (*Halbgruppenfahrzeug*), technical support vehicle (*Bergepanzerwagen 38 D*) and self-propelled anti-aircraft gun armed with twin 3-cm *MK 103* autocannons (*Kugelblitz 38 D*). None of these projects reached the stage of a wooden mockup, not to mention a fully-fledged prototype.

Combat history

The *Jagdpanzer 38* was looked upon by the Germans as the direct successor of the *Marder III* tank destroyer, so according to original assumptions they were to be supplied to the same tank hunter units (*Panzerjäger-Abteilung*) which operated the *Marders*. However in early 1944 the plans were changed and it was decided to allot the Hetzers to tank hunter units, but these units would be assigned only to infantry divisions and inde-

pendent tank hunter battalions (*Armee-Panzerjäger-Abteilung*). So in contrast with the *Marder IIIs* the *Hetzers* were not to be assigned to armored divisions.

Early decisions assumed that the tank hunter companies armed with *Jagdpanzer 38s* would have been formed on the basis of organizational schemes, on which analogical units, equipped with *Marder IIIs* were previously formed. In 1944, due to changes of nomenclature of some units equipped with armored vehicles the tank hunter company began to be formally referred to as assault gun company and it was decided that it would be formed in accordance with the newly created *K.St.N. 1149* organizational scheme (carried into effect on 1 February 1944), existing in two variants: model A (*Ausführung A*) and model B (*Ausführung B*). The difference between the two "versions" of the organizational scheme consisted in the number of the vehicles in the company: according to *K.St.N. 1149 Ausführung A* the company numbered in total 118 soldiers and 10 vehicles while under *K.St.N. 1149 Ausführung B* – probably 134 soldiers and 14 vehicles. According

14 mm Flammenwerfer 41 flame thrower installed instead of *7,5 cm PaK 39 L/48* gun in *Flammpanzer 38(t) Hetzers*. The tube itself was protected by an exterior barrel connected with the mantlet. 700 liters of the incendiary-compound preserved in the interior reservoir was enough for 24 single shots or one lasting 90 s volley on the 60 m range. [Kagero Archive]

to current knowledge, the *K.St.N. 1149* organizational scheme was modified at least once, moreover – from approximately third quarter of 1944 the assault gun companies equipped with *Hetzers* and assigned to infantry divisions began to be referred to as tank hunter companies again.

Due to the necessity of making the soldiers acquainted with the design and characteristics, conducting necessary test and procuring the maintenance documentations for the mechanics the first produced *Jagdpanzer 38s* were assigned to training centers in Kummersdorf (five vehicles), Berka (three vehicles), Bergen (two vehicles) or Hillersleben. Tests at proving ground ended with good results however the vehicle had several drawbacks. The main disadvantage was the aforementioned overloading of the forward section of the chassis resulting in "nose droop". This

ostensibly minor flaw implicated many unfavorable effects, concerning mainly the aiming. There were also reservations about the tracks which were not enough wide and the idler wheels. Disastrous situation on the Eastern Front resulting from the Operation "Bagration" commenced in June of 1944 excluded the possibility of implementation of a repair plan which would eliminate even a few flaws and forced the introduction of the vehicle to frontline service as quickly as it was possible. The elimination of the design failures was to be conducted during the production. The first combat unit to be equipped with the *Hetzers* was the *Heeres-Panzerjäger-Abteilung 731*, which received 45 vehicles (it consisted of three companies with 14 vehicles in each and three additional destroyers in the headquarters) during 4 -13 July 1944. This unit was sent to support the Army Group "North".

Flammpanzer 38(t) Hetzer captured by US Soldiers during the Ardennes offensive in the first days of January 1945. Twenty self-propelled flame throwers, armed by *BMM*, had been attached to *Panzer-Flamm-Kompanie 352* and *Panzer-Flamm-Kompanie 353*, assigned to the Army Group G. The photographed vehicle has the three-tone camouflage scheme applied in *Škoda* factory, consisted of vertical, sharp-edged chocolate brown (RAL 8017), dark green (RAL 6003) and sand yellow (RAL 7028) spots. [Kagero Archive]

A pair of *Jagpdanzer 38 (t) Hetzers* destroyed in one of the German towns in the first weeks of 1945. The forepart vehicle has the road wheels with narrower rubber rims, indicating that it had been completed during last war-period autumn. The *Kugellafette V* gun mantlet is also clearly visible. The three-tone camouflage scheme consisted of wide dark green (RAL 6003) and chocolate brown (RAL 8017) spots on the sand yellow (RAL 7028) background, which could be seen on the front armor plate suggests that it had been manufactured in *BMM*. Both of them were destroyed by the internal explosion, probably initiated by retreating crews. [Kagero Archive]

Jagdpanzer 38(t) Hetzer disrupted probably by the interior shells explosion. It has the *Kugellafette V* gun mantlet. The barely seen camouflage scheme had been applied just after the end of manufacturing in *BMM* factory and represents the first version employed by the Prague works – wide dark green (RAL 6003) and chocolate brown (RAL 8017) spots over the sand yellow (RAL 7028) background. There is a white national emblem on the superstructure side – the contoured cross with short arms. The western front, 1944/1945. [Kagero Archive]

One of twenty self-propelled flame throwers Flammpanzer 38(t) Hetzer, constructed in December 1944 by *BMM*, captured by the US Army in the first days of January 1945. The converted gun mantlet and barrel are clearly visible. The vehicle has the first type of the three-tone camouflage scheme – wide dark green (RAL 6003) and chocolate brown (RAL 8017) spots over the sand yellow (RAL 7028) background, applied in the production halls of the Prague factory. [Kagero Archive]

Another unit equipped with the *Jagdpanzer 38s* was the *Heeres-Panzerjäger-Abteilung 743* which due to swift destruction of the Army Group "Centre" was sent to support this group. Like the predecessor the unit received 45 Hetzers during 19-28 July 1944. Noteworthy is the fact that elements of the *Heeres-Panzerjäger-Abteilung 743* in the strength of 1. and 2. companies took part in the pacification of the Warsaw Uprising, loosing one vehicle on 2 August (belonging to *2./H.Pz.Jg.Abt. 743*) which was attacked by the insurgents with grenades and Molotov cocktails at Szpitalna street and then captured. This success is attributed to soldiers of the Kedyw "Kolegium C" company and 1. and 2. companies of the "Kiliński" battalion. On the evening of that day the captured *Hetzer* was used as the strengthening element of the barricade built between Bouden and Sienkiewicz streets, fencing off the Napoleon Square from Szpitalna street. Although the damage of the vehicle was estimated at 70%, in mid August the insurgents managed to bring it back to use, however it did not see combat since traveling through Warsaw arteries would require disman-

tling of some barricades, which was not allowed. The "Chwat" ("Merry Blade" – this was the name given to the captured *Hetzer*) was placed in the building of the Main Post Office and served as a reserve weapon until this building was bombed by the Germans and collapsed. The vehicle remained under the rubble till next year. The *Heeres-Panzerjäger-Abteilung 743* suffered further losses on 9 or 10 August (one vehicle totally destroyed) and on the last day of its commitment in the fighting with the Poles, i.e. on 14 August (four vehicles heavily damaged).

In late summer of 1944 the *Jagdpanzer 38s* began to reinforce units fighting in the main trouble spots of the front line or in places where a knocking-out Soviet offensive was expected. In August a company of *Hetzers* was assigned to the 15th and 76th Infantry Divisions and *8. SS-Kavallerie-Division „Florian Geyer".* Several analogical units, which ultimately were not permanently assigned to any infantry division, since they were entrusted with the role of independent tactical formations supporting various units, depending on the situation on the

Coated with thick layer of snow, abandoned *Jagdpanzer 38(t) Hetzer* representing the late production series. The *Kugellafette V* gun mantlet, road wheels with narrower rubber rims and the fourth type of idler wheel with six wide circular holes are clearly visible. The vehicle has probably the standard, three-tone camouflage applied by the crews themselves. The eastern front, 1944/1945. [Kagero Archive]

Jagdpanzer 38(t) Hetzer attached do *Panzergruppe Beyer* during the battle of Kolberg in march 1945. The vehicle represents the early production series which suggests the first type of muffler with the pierced extra-protection plate and lack of side skirts alongside the superstructure edges. There is a flag of the III Reich installed on the rear plate, probably used as a quick-recognize mark for the German forces. The two-tone camouflage scheme recoins from the remnants of the white-wash paint. Lack of the essential damages and the open hatches suggest that it had not been destroyed but rather abandoned. [Museum of the Polish Arms Archive]

front line, were also formed. A month later a small number of companies equipped with Hetzers, with one, incomplete *Heeres-Panzerjäger-Abteilung* (number 741) was sent to the Western Front. The latter unit took part in operations near Arnhem, the objective of which was the annihilation of the Allied operation codenamed "Market-Garden".

Between October and December of 1944 the Hetzers were assigned among others to following units of the German army: 1009 (9. Volksgrenadier Division), 1026 (Volksgrenadier Division), 1044 (44. Infantry Division), 1179 (257. Volksgrenadier Division), 1222 (181. Infantry Division), 1147 (47. Volksgrenadier Division), 1162 (62. Volksgrenadier Division), 1167 (167. Volksgrenadier Division), 1168 (68. Infantry Division), 1194 (94. Infantry Division; in December of 1944, when the *Hetzers* arrived, it was fighting in Italy), 1272 (272. Volksgrenadier Division), 1304 (304. Infantry Division), 1316 (16.

Immobilized *Jagdpanzer 38(t) Hetzer* of *Panzergruppe Beyer*, solidly installed in one of the barricades. The *Kugellafette V* gun mantlet and the case mounted on the front part of right fender are clearly visible. There are remnants of additional branch-camouflage on the armor plates which is quite interesting as far as the street fights are concerned. The vehicle is probably painted in the standard three-tone scheme. Kolberg, March 1945. [Museum of the Polish Arms Archive]

Jagdpanzer 38(t) Hetzer abandoned in the vicinity of Budapest in February 1945. The camouflage scheme represents the first type of coat factory-applied in *BMM*. There is white three digit, so called *wreck number* usually applied by the Soviets on captured vehicles visible on the superstructure side as well as an arrow marking the large hole in the gun mantlet base. [Kagero Archive]

Volksgrenadier Division), 1320 (320. Volksgrenadier Division), 1326 (326. Volksgrenadier Division), 1337 (337. Volksgrenadier Division), 1340 (340. Volksgrenadier Division), 1349 (349. Volksgrenadier Division), 1352 (352. Volksgrenadier Division), 1363 (363. Volksgrenadier Division), 1510 (73. Infantry Division), 1560 (560. Volksgrenadier Division), 1708 (17. Volksgrenadier Division), 1711 (711. Infantry Division), 1716 (716. Volksgrenadier Division) and 1818 (18. Volksgrenadier Division).

The year 1945 began tragically for the Germans. In the West offensive actions conducted under the Operation "Wacht am Rhein" were stopped and in the East the Red Army launched a new, massive assault intended to reach the gates of Berlin. To stop advancing enemies the Germans started hastily forming new units, including companies equipped with *Hetzers*, which began to be assigned to provisional combat formations, like the Panzergrenadier Division "Kurmark" (1551 Company) or the Armored Division *Feldhernhalle*. Units of new organizational

Jagdpanzer 38(t) Hetzer abandoned on the roadside somewhere in Sudeten, April 1945. The vehicle had been completed between October and November 1944 and had the late version of the *Kugellafette V* gun mantlet. The camouflage scheme represents the first type of coat factory-applied in *BMM*, consisted of irregular, wide, vertical chocolate brown (RAL 8012) and dark green (RAL 6003) spots on the sand yellow (RAL 7028) background. [Kagero Archive]

Jagdpanzer 38(t) Hetzer "Renate Boss" destroyed in the area of Gdańsk (Danzig), March 1945. The vehicle was probably manufactured in autumn of the previous year. Its nickname looks to be written in black over white background. [Kagero Archive]

structure were also formed (so-called tank hunter brigades; *Panzerjäger-Brigade*); one of these units, number 104 consisted of as many as six companies with 84 vehicles. The *Jagdpanzer 38s* also found their way to units, in which they theoretically should not be, namely the 507. i 511. *s.H.Pz.Abt* (heavy tank battalions). It was caused by the lack of the Tiger tanks being the main equipment of units of this type.

It is worth mentioning that the Hetzers did not serve exclusively in the Wehrmacht and Waffen-SS. In December of 1944 and January of 1945 the Hungarian Army received 75 vehicles of this type. They reportedly earned much better reputation among the soldiers than the *Zrínyi* self-propelled guns, which should not surprise, taking into consideration the disproportions in armament. Several *Jagdpanzer 38s* were supplied to

Late production series *Jagdpanzer 38(t) Hetzer* destroyed by the US Army in West Germany, spring 1945. The vehicle has the road wheels with narrower rubber rims introduced in August 1944, and the second type of idler wheels with seven circular holes and radial ribs. The gun mantlet represented the late version of *Kugellafette V*. The two-tone camouflage scheme consisted of sharp-edged dark green (RAL 6003) spots on the sand yellow (RAL 7028) background. It was probably factory-applied by Škoda. There is also the national mark – the contoured, short-armed cross and the part of tactical number – 32. Both elements painted in white. [Kagero Archive]

Jagdpanzer 38(t) Hetzer manufactured by *Škoda*, destroyed somewhere in Bohemia in the beginning of 1945. The camouflage scheme represents the characteristic for the factory in Pilsne variant, consisted of vertical dark green (RAL 6003), chocolate brown (RAL 8017) and plain sand yellow (RAL 7028). Black or dark red tactical number – 201 – is clearly visible too. [Kagero Archive]

Jagdpanzer 38(t) Hetzer with unusual T-033 code number, applied probably by the Soviets after the vehicle had been ca abandoned by its crew in Hungary, in the last stage of World War II. Note the late version of *Kugellafette V* gun mantlet. [Kagero Archive]

Bohemian civilians riding *Jagdpanzer 38(t) Hetzer* abandoned by its crew somewhere in the nowadays Czech Republic. The vehicle's camouflage scheme is supplemented by cut down branches. May 1945. [Kagero Archive]

Jagdpanzer 38(t) Hetzer totally destroyed during the Prague uprising in May 1945. The vehicle represents the late production variant with narrower rubber wheel rims and *Kugellafette V* gun mantlet. [Kagero Archive]

Late production series *Jagdpanzer 38(t) Hetzer "Thule"* destroyed in Berlin, April 1945. The large nickname painted on the superstructure side was very uncommon. Usually the names were placed on the front plate. [Kagero Archive]

Jagdpanzer 38(t) Hetzer destroyed in the Budapest area, February 1945. The white 21 so called wreck number and hit traces encirclements were painted by the Soviets after the vehicle had been captured. [Kagero Archive]

Jagdpanzer 38(t) Hetzer **with *BMM* factory-applied camouflage scheme destroyed in the vicinity of Budapest in February 1945. The white 115 so called wreck number was applied after the vehicle had been captured by the Soviets. [Kagero Archive]**

ROA (Russian Liberation Army) units. Approximate assessments allow conclusion that their number was not smaller than 10 and not larger than 20. In March and April of 1945 three *Hetzers*, chassis numbers 322549, 323329 and 323358 were assigned to the 5th self-propelled gun squadron of the 6th Infantry Division of the Polish People's Army. In the unit's documentation their German designation was replaced by the term "self-propelled gun T-38 (75 mm)".

In the conclusion it is worth mentioning that the history of the *Hetzer* did not end with the surrender of the Third Reich. Immediately after the war the *Škoda* works resumed production of the *Jagdpanzer 38s* for the Czechoslovak army, using components ready for assembly. Slightly modified vehicles (with the machine gun position removed) entered service under the designation *ST-1*. The other and last post-war purchaser of the *Hetzers* was the Swiss army, which received them from the Czech manufacturer under the designation *G-13*. The delivered vehicles differed from the original primarily in the armament (the *7,5 cm Pak 39 L/48* gun was replaced by the *7,5 cm Stu.K 40* with muzzle brake) and the presence of the evacuation hatch in the floor, near the gunner position. During the service the original petrol engines were replaced by *Saurer-Arbon CH-2DRM* Diesel engines. The G-13 vehicles remained in Swiss armed forces service until the turn of 1971/72.

Jagdpanzer 38(t) Hetzer **manufactured probably by *Škoda*, destroyed west of Berlin, May 1945. The camouflage scheme represents the characteristic for this factory type, consisted of vertical dark green (RAL 6003), chocolate brown (RAL 8017) and plain sand yellow (RAL 7028). [Kagero Archive]**

Jagdpanzer 38(t) Hetzer **destroyed in Gdańsk (Danzig) in March 1945. There are remnants of white-wash camouflage paint visible on the superstructure as well as steel helmets of the vehicle's crew installed on the side plate. [Kagero Archive]**

Panzerjäger 38(t) Hetzer & G-13

Jagdpanzer 38(t) Hetzer "Steffi" captured probably by the Prague insurgents in May 1945. The nickname painted in white on the gun mantlet base is clearly visible. [Kagero Archive]

Rear view of *Jagdpanzer 38(t) Hetzer* with code number T-038 captured by the Soviets during the Hungarian campaign, February 1945. Spare track links and muffler are clearly visible. [Kagero Archive]

Another *Jagdpanzer 38(t) Hetzer* bearing the *"Marika"* nickname, abandoned by its German Crew and captured by the Soviets in Hungary, the beginning of 1945. The white 38 wreck number was painted after the vehicle had been lost. [Kagero Archive]

Jagdpanzer 38(t) Hetzer abandoned during the Prague uprising in May 1945. The vehicle bears double tactical number: 153 painted over previous 003. The gun mantlet seems to be missed. [Kagero Archive]

Jagdpanzer 38(t) Hetzer of the German *320. Panzer-Division* destroyed in the vicinity of Ostrava in May 1945. The vehicle's camouflage consists of large probably dark green (RAL 6003) spots over the sand yellow (RAL 7028) background and was applied by the crew itself. [Kagero Archive]

Chwat

One of the best known Jagdpanzer 38(t) Hetzer tank destroyers in Poland is the exemplar captured on August 2, 1944 during the Warsaw Uprising. The original users of this vehicle were soldiers of the German Panzerjäger Abteilung 743 (743rd Armored Tank Hunters Battalion). On the morning of the second day of the Polish uprising a group of the unit's combat vehicles was deployed to the Napoleon Square area, in support of the Wehrmacht soldiers trying to re-take the Main Post Office building. One of the machines manufactured by the BMM factory (chassis number 321078) was attacked by insurgents using hand grenades and bottles with fuel, so-called Molotov cocktails. As a result of the fire its crew abandoned the vehicle, and it itself, partially burned, fell in Polish hands.

Shortly thereafter, it was incorporated into the Napoleon Square barricade. On August 5 it was withdrawn to the Main Post Office building for repair. Despite the fact that the vehicle had been destroyed in approximately 70 percent, by August 10 it was restored to the running condition. In honor of the soldiers of the Polish Home Army Headquarters Propaganda Department Protection Unit nicknamed „Chwaty" who conducted the repair, the captured Hetzer received the name "Chwat" ("Merry Blade"). The white eagle national emblem and the name of the machine were painted by the Polish artist Leon Michalski.

Because of the difficulty in moving through barricades, the Polish manned Hetzer did not take part in the later part of the battle of Warsaw. On September 5 it was buried under the rubble of the Main Post Office, which was destroyed by a German air raid and rocket artillery fire. Excavated a year after the surrender of the Third Reich for the next several years it was exhibited in the Polish Army Museum in Warsaw. According to the decision of the Central Political Board of the Polish Army, it was scrapped in 1950. Only one original wheel of the "Chwat" survived until present. From the beginning of 21st century it is exhibited in the main military museum in the capital of Poland.

Jagdpanzer 38(t) Hetzer "Chwat" ("Merry Blade") **captured by the Warsaw insurgents on August 2, 1944. Despite its battle damage the vehicle were brought back to the running condition. [Kagero Archive]**

which the Germans got the opportunity to saturate their units (mainly of the infantry) with anti-tank weapons. Observing the role played by these destroyers in the final period of the war there is no exaggeration in the conclusion that at the end of the war the *Jagdpanzer 38s* belonged to the group of the most important German fighting vehicles, frequently delivering the endangered front lines from breaching in several sections.

Summary

Despite their disadvantages the *Hetzer* tank destroyers were undoubtedly among the best German fighting vehicles designed during the war. Satisfactory mobility, comparatively powerful armament and pretty thick and well profiled armor allowed them to fight on equal terms with most enemy vehicles, excluding the heaviest ones (*IS-2, ISU-152, M-26 Pershing* etc.). These assets were additionally supplemented by small size of the vehicle, making it a difficult target, even at close distances. The simplicity of the design and ease of production resulted in the possibility of producing them in great numbers, thanks to

Bibliography

Books and articles:

Magnuski Janusz, *Wozy bojowe LWP 1943 – 1983*, Warsaw 1985.

Thomas L. Jentz, Doyle L. Hilary, Sarson Peter, *Flammpanzer. German flamethrowers 1941 – 1945*, London 1995.

Ledwoch Janusz, *Jagdpanzer 38(t) "Hetzer"*, part 1, Warsaw 1997.

Thomas L. Jentz, Doyle L. Hilary, Badrocke Mike, *Jagdpanzer 38 'Hetzer' 1944–1945*, Oxford 2001.

Thomas L. Jentz, Doyle L. Hilary, *Artillerie Selbstfahrlafetten*, Panzer Tracts No. 10, Boyds 2002.

Thomas L. Jentz, Doyle L. Hilary, *Paper Panzers*, Panzer Tracts No. 20-1, Boyds 2002.

Thomas L. Jentz, Doyle L. Hilary, *Paper Panzers*, Panzer Tracts No. 20-2, Boyds 2002.

Thomas L. Jentz, Doyle L. Hilary, *Bergerpanzerwagen*, Panzer Tracts No. 16, Boyds 2004.

Frances Vladimír, Miroslav Bíly, *Jagdpanzer 38 Hetzer*, Prague 2006.

Spielberger J. Walter, Doyle L. Hilary, Jentz L. Thomas, *Light Jagdpanzer*, Atglen 2007.

Eugeniusz Żygulski, *Panzerjäger 38(T) Hetzer – zwiastun klęski*, Technika Wojskowa Historia, special issue 2/2013, pp. 24-36.

Jagdpanzer 38(t) Hetzer "Chwat" ("Merry Blade") **captured by the Poles during the Warsaw uprising in August 1944. Formerly the vehicle belonged to** *Panzerjäger Abteilung 743* **which tank destroyers were sent to support Wehrmacht during the assault on the Main Office Building. Here the** *"Chwat"* **is seen at the corner of Boudena and Szpitalna streets as an element of the barricade. [Kagero Archive]**

Jagdpanzer 38(t) Hetzer „Chwat" („Merry Blade") after being fully restored by the Polish insurgents. The national markings and vehicle's nickname are clearly visible Warsaw, August 1944. [Kagero Archive]

Hetzer of the „Hermann von Salza" regiment, outskirts of Berlin, April 1945

At the moment when the Third Reich fell into the ultimate collapse, eyes of most German military commanders and their Allied opponents were set on Berlin, where the relentless struggle between the city and garrison units, which managed to break through to it and the Red Army was raging.

One of the units, which took its combat position on the outskirts of the capital of Adolf Hitler's state in the second half of April 1945 was the SS-Panzer-Abteilung 11 "Hermann von Salza" (named after the famous Grand Master of the Teutonic Order of the 13th century and raised to the status of Armored Regiment as late as 11 April). This unit, composed of two armored battalions (commanded by SS-Sturmbannführer Gratwohl and SS-Sturmbannführer Hertzig), led at the time by the SS-Obersturmbannführer Kausch, supported by the soldiers of the SS-Pionier-Battalion 11 took the brunt of the Soviet impact in the area of Reichenow. On 19 April it was forced to withdraw towards Strausberg. At this time, the armored strength of the regiment comprised several Pz.Kpfw. IV and Pz.Kpfw. V Panther medium tanks, 50 StuG III Ausf. G assault guns and 10 tracked tank destroyers – Jagdpanzer 38 (t) Hetzer.

The photo session, presented on the next page, was completed in June 2012 by the members of the Polish "Die Freiwilligen" Reenactment Association. It was an attempt to re-enact the last days of fighting of the Jag panzer 38 (t) Hetzer crew of the SS armored regiment mentioned above, somewhere on the outskirts of Berlin. Uniforms and equipment used in its implementation is an accurate representation of sets used by the crew of such

vehicles, typically assigned to tank units; soldiers are dressed in special uniforms made of black cloth, not, as it was in the case of self-propelled artillery or assault artillery - gray-green (Feldgrau). They are complemented by denim elements in Erbsentarnmuster 44 camouflage pattern, commonly used in the Waffen-SS from the summer of 1944.

The vehicle involved in the photo shoot was the late production series Jag panzer 38(t) Hetzer, belonging to the Handmet Military collection in Gostyń (Poland). The machine was originally drawn into the Swiss army arsenal operating the G-13 variant and remained in service until the seventies. After the transfer to Handmet Military collection it has undergone a thorough renovation, during which the signs of the Third Reich military office (Waffenamt) placed among other things on the main armament were discovered. The owner also managed to read the armor casting date - October 21, 1944, as well as its manufacturer, which was Poldi Hűtte in Kladno, Czech Republic.

Late provenance of the vehicle is also confirmed by the machine construction details. The gun mantlet represents the late variant: Kugellafette V. In addition, the narrower rubber bandages of road wheels and the idler wheel equipped with four large weight-reduction holes also represent production from the last period of existence of the Third Reich. Following this the Hetzer from Gostyń received universal, sand yellow monochromatic painting common during the years 1944-1945. The three-digit tactical number is also typical of those used in the ranks of the German armored units in the second half of World War Two.

The roadworthy vehicle's debut took place during the Podrzecze (Poland) "Military Zone" reenactment meeting in July 2011. During the next edition of this event it was displayed

to the general public in all its glory, with a machine gun. The complete crew in historical uniforms, performed refueling, tapping track pins and ammunition loading work in front of the audience Today it is an object of interest in people involved in a kind of historical reconstruction and embellishment of the

Handmet Military collection. It is the only roadworthy vehicle of this type in Poland and one of few worldwide.

Łukasz Gładysiak
Photos by Łukasz Dyczkowski, www.fotogeniczny.pl.

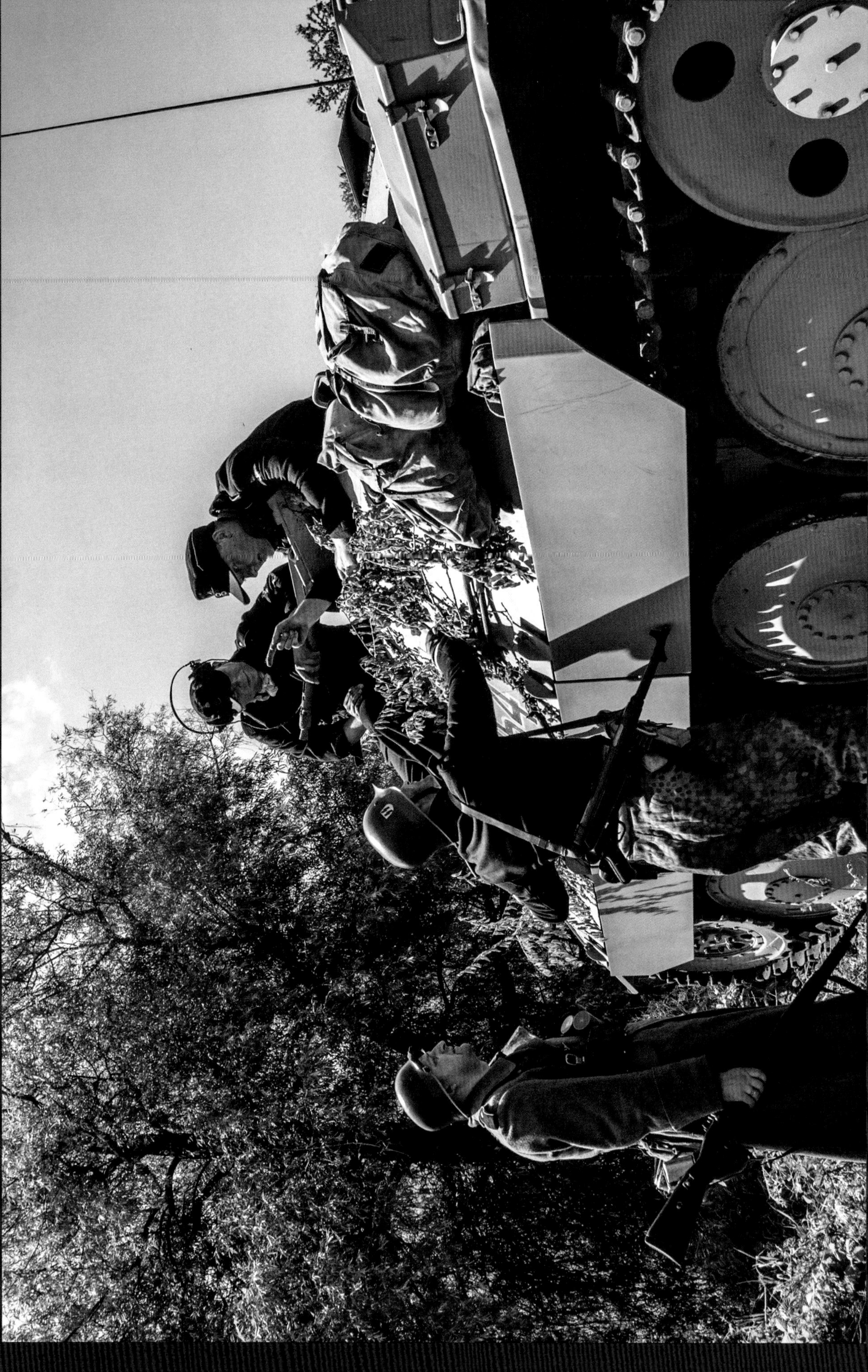

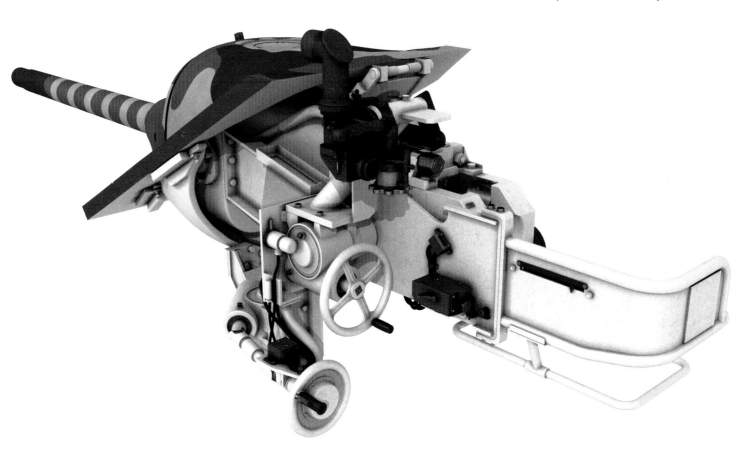

7,5 cm PaK 39 gun – the main armament of *Jagdpanzer 38(t) Hetzer,* left side seen from two different angles. The *Sfl ZF1a* telescopic sight is also clearly visible.

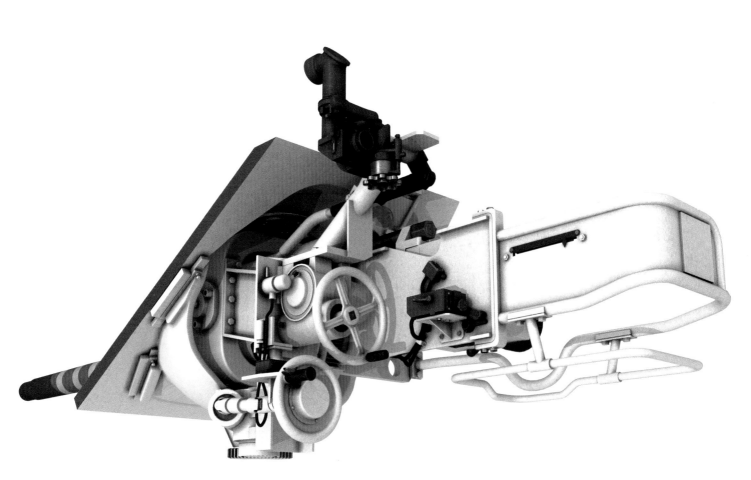

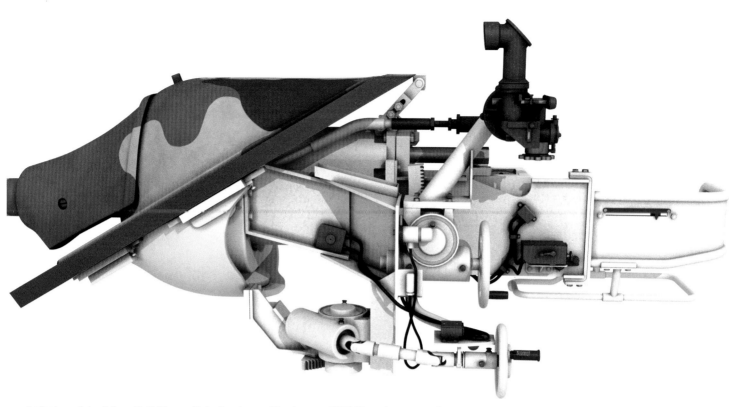

Left view of the *7,5 cm PaK 39* gun. Note the shape of *Jagdpanzer 38(t) Hetzer*'s gun mantlet.

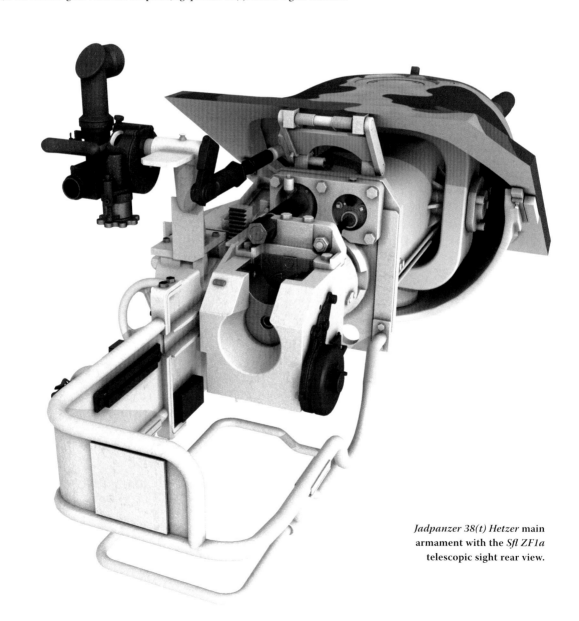

Jadpanzer 38(t) Hetzer **main armament with the** *Sfl ZF1a* **telescopic sight rear view.**

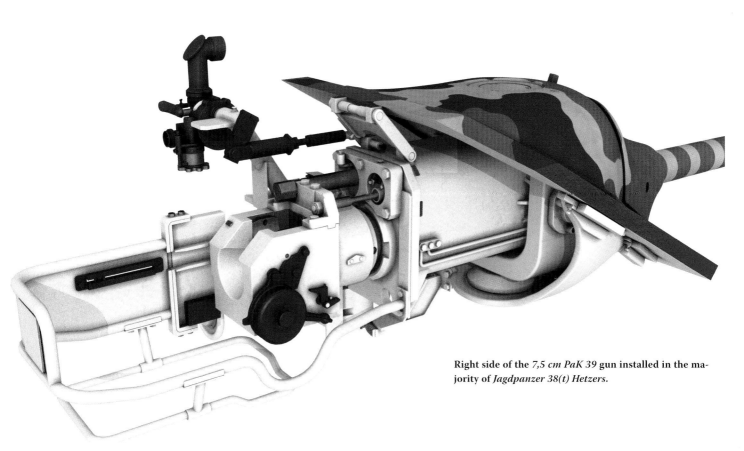

Right side of the *7,5 cm PaK 39* gun installed in the majority of *Jagdpanzer 38(t) Hetzers.*

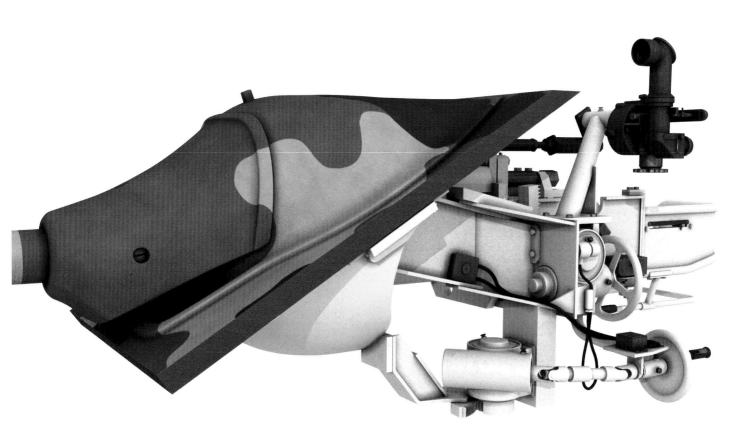

7,5 cm PaK 39 gun and the *Saukopfsblende* (the *boar head*) gun mantlet seen from the left front corner of the vehicle.

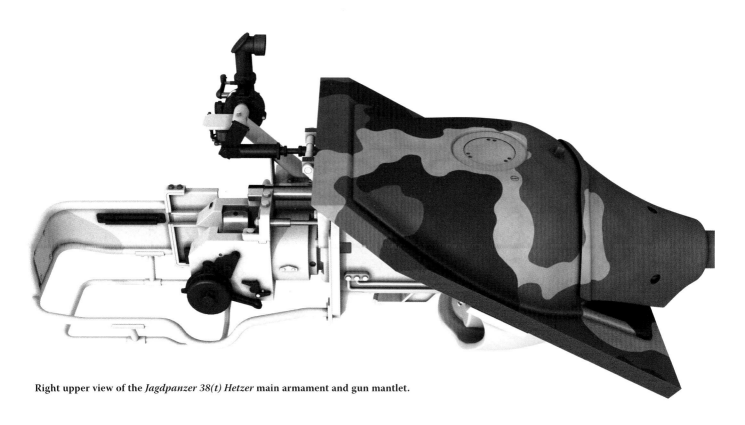

Right upper view of the *Jagdpanzer 38(t) Hetzer* main armament and gun mantlet.

7,5 cm PaK 39 gun seen from behind. Note the tubular shell protective shield

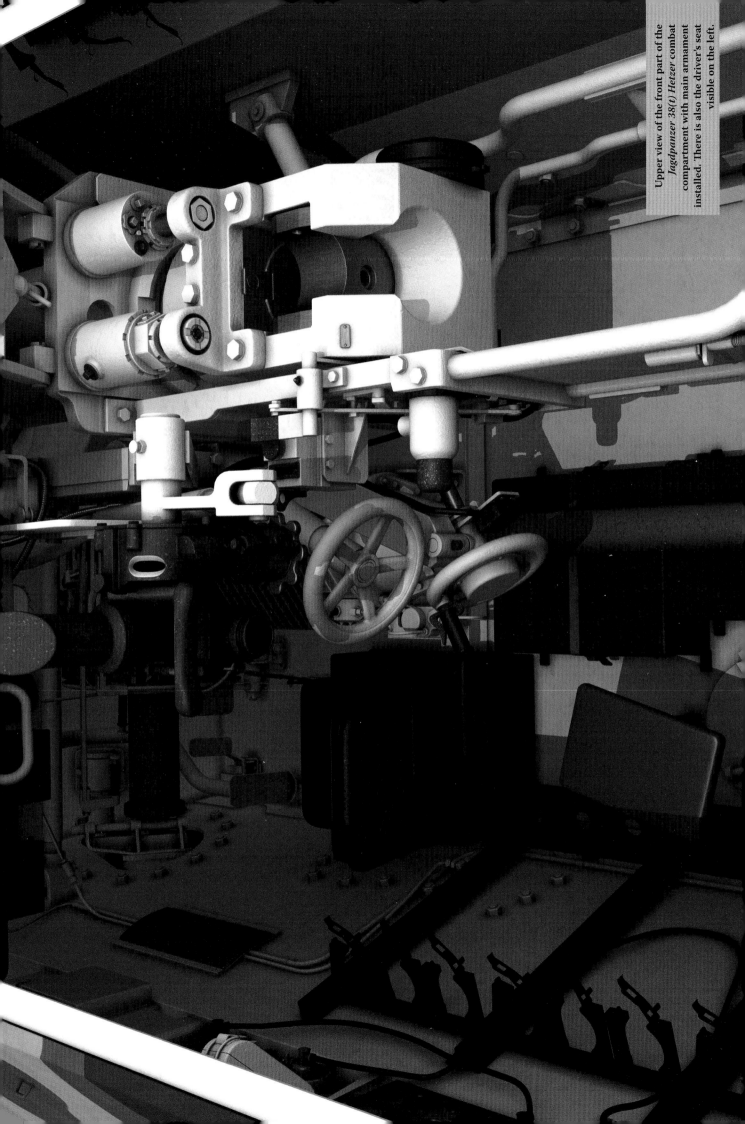

160 -IP, liquid-cooled *Praga AC 2800* petrol engine – the powerplant of *Jagdpanzer 38(t) Hetzer.*

Upper view of the left part of the *Jagdpanzer 38(t) Hetzer* engine compartment. The fuel tank and battery are clearly visible.

Another view of the *Jagdpanzer 38(t) Hetzer* engine compartment with a battery and fuel tank.

Praga AC 2800 engine filters installed by the right wall of the Jagdpanzer 38(t) Hetzer engine compartment.

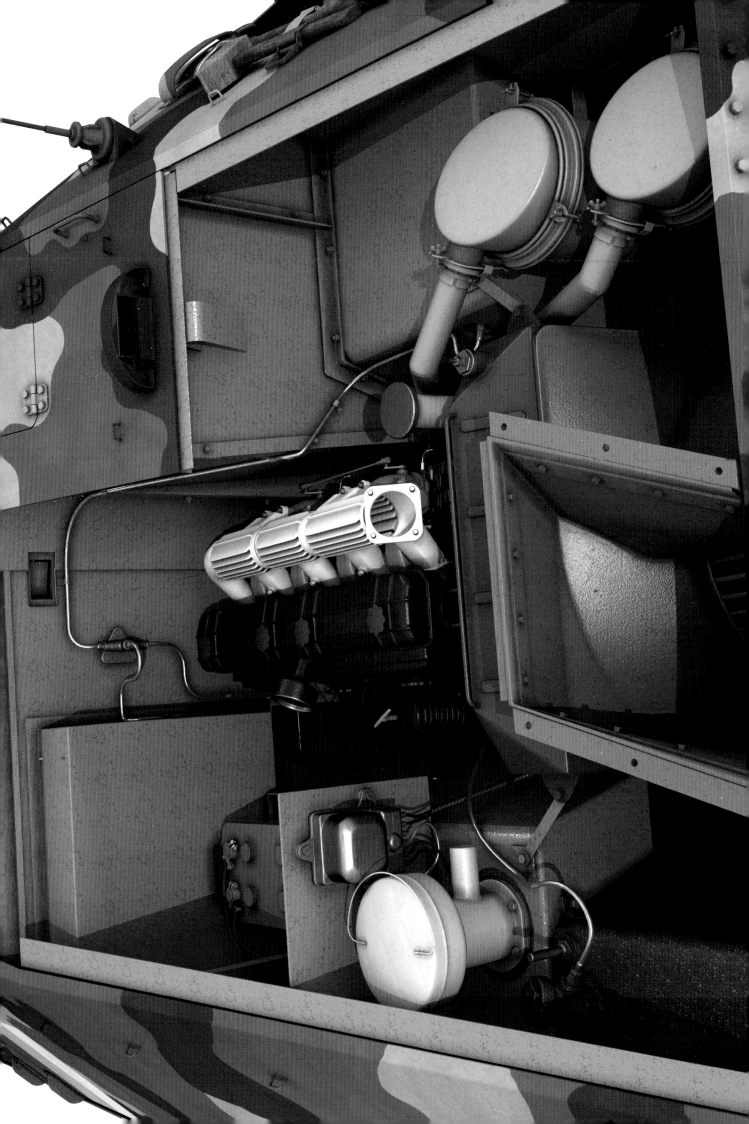

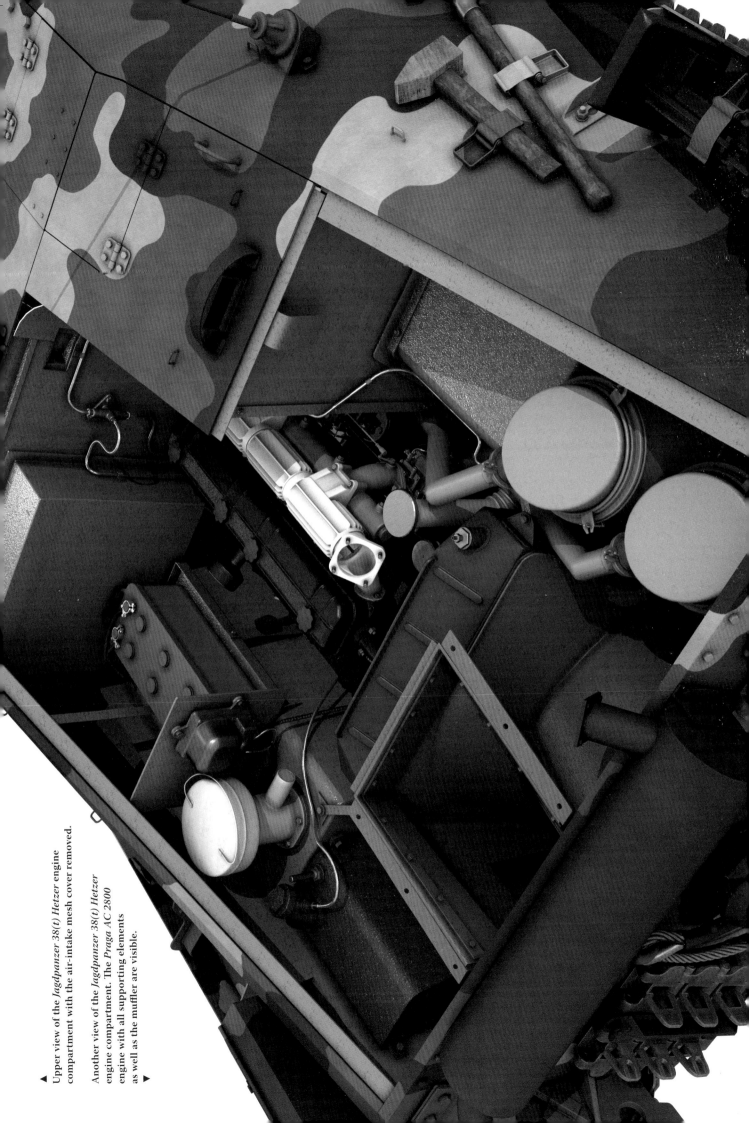

Upper view of the *Jagdpanzer 38(t) Hetzer* engine compartment with the air-intake mesh cover removed. ◄

Another view of the *Jagdpanzer 38(t) Hetzer* engine compartment. The *Praga AC 2800* engine with all supporting elements as well as the muffler are visible. ►

Praga AC 2800 engine with filters, battery and air-intake box without its cover seen from above.

The powerplant of *Jagdpanzer 38(t) Hetzer* was the Czech-made *Praga AC 2800* 160 HP petrol engine. Note the details of this 7754 ccm device on this and another two pages.

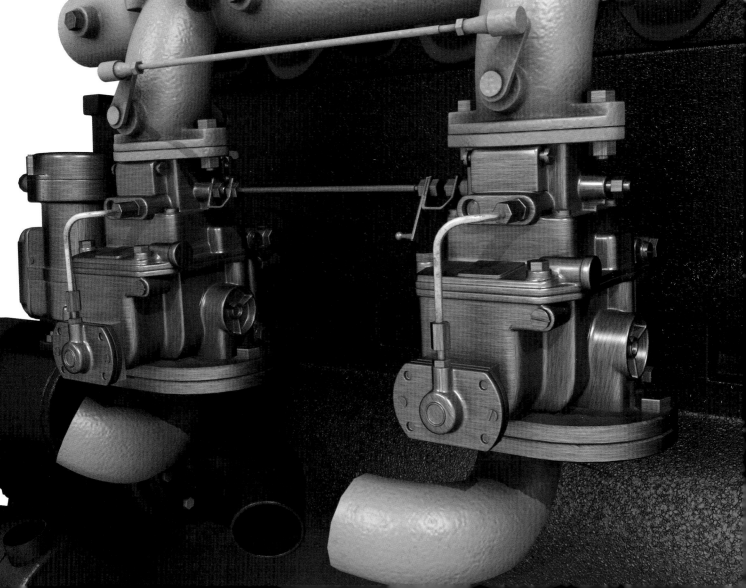

Left idler and the last road wheel of *Jagdpanzer 38(t) Hetzer* with tracks installed.

Rear left trolley of the *Jagdpanzer 38(t) Hetzer* suspension seen from below. The road wheels are installed.

The first left trolley of the *Jagdpanzer 38(t) Hetzer* suspension with one of the road wheels removed. Note the massive leaf spring and a part of drive sprocket.

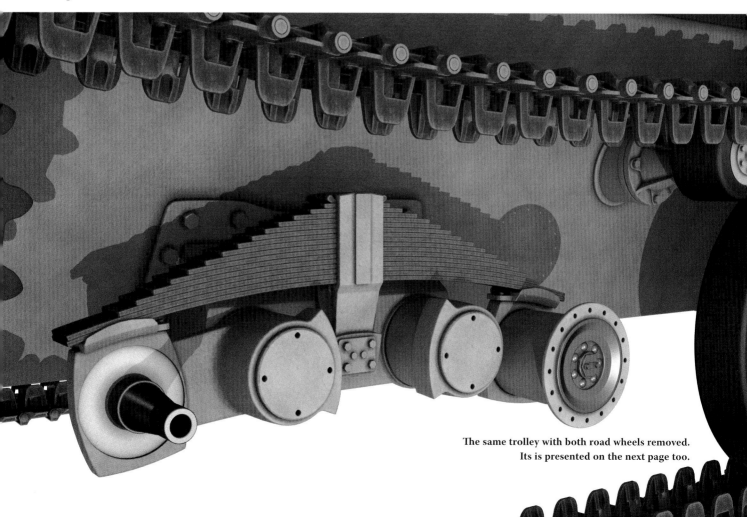

The same trolley with both road wheels removed.
Its is presented on the next page too.

The first left *Jagdpanzer 38(t) Hetzer* trolley mount without road wheel arms and leaf spring.

Jagdpanzer 38(t) Hetzer right idler wheel.

Jagdpanzer 38(t) Hetzer right idler wheel axle and inner mechanism bolted cover.

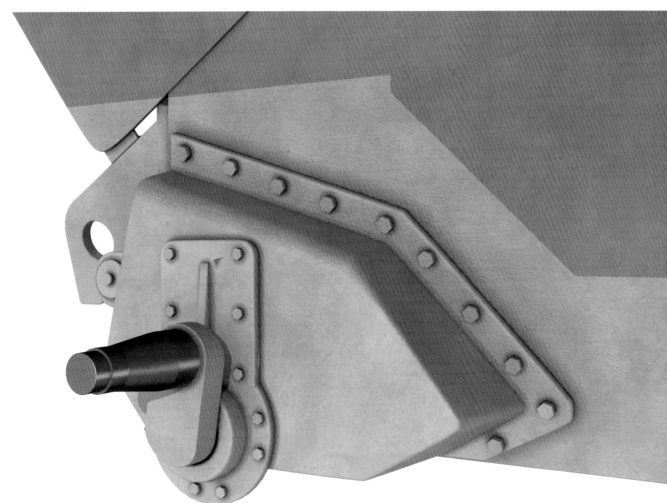

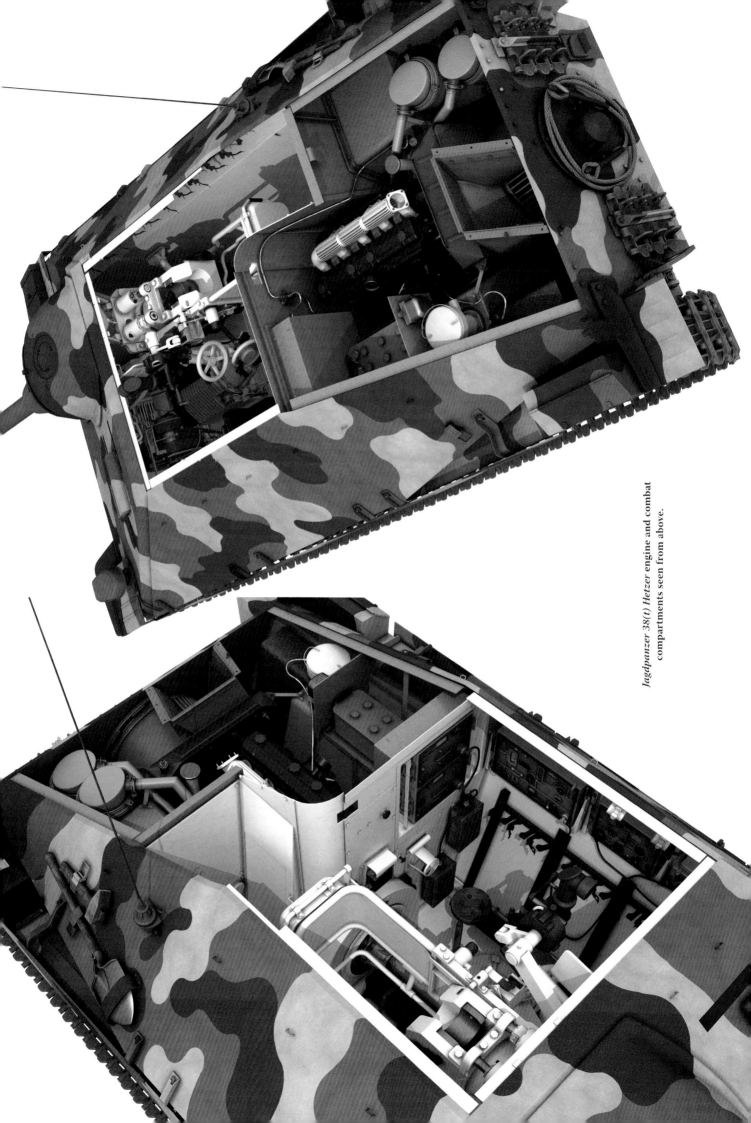

Jagdpanzer 38(t) Hetzer engine and combat compartments seen from above.

Upper view of the *Jagdpanzer 38(t) Hetzer* tank destroyer. The camouflage scheme represents one of the factory-applied, late war variations.

Front view of *Jagdpanzer 38(t) Hetzer*. The 75 mm gun barrel with the *Saukopfsblende* mantlet are clearly visible.

Right corner view of *Jagdpanzer 38(t) Hetzer*. Note the external stowage and characteristic, horizontally mounted muffler.

Upper view of *Jagdpanzer 38(t) Hetzer*.
The large engine hatch is open.

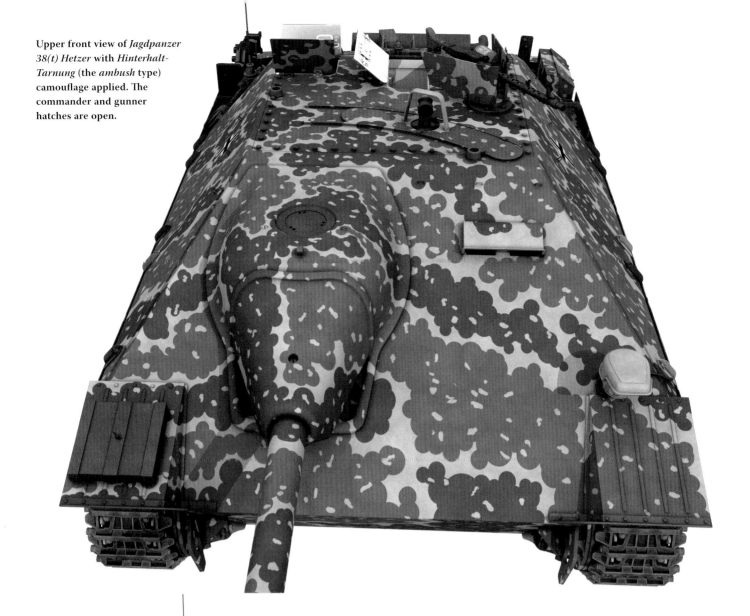

Upper front view of *Jagdpanzer 38(t) Hetzer* with *Hinterhalt-Tarnung* (the *ambush* type) camouflage applied. The commander and gunner hatches are open.

Right side of the same vehicle. There are wooden track blocks installed on the front fender.

Commander and gunner hatches of *Jagdpanzer 38(t) Hetzer* in open positions. The inside-operated *7,92 mm MG 34* machine gun is visible too.

Another view of the upper part of the *Jagdpanzer 38(t) Hetzer* superstructure with *7,92 mm MG 34* machine gun and *Sfl ZF1a* telescopis sight. There are also two jib sockets visible.

Left front corner of *Jagdpanzer 38(t) Hetzer* with the *Notek* light installed on the fender.

Commander and gunner hatches of *Jagdpanzer 38(t) Hetzer* in open position seen from behind.

Rear wall of the *Jagdpanzer 38(t)*
Hetzer superstructure with muffler,
spare track links and towing cable
installed.

Jagdpanzer 38(t) Hetzer muffler and
air-intake mesh cover seen from
above. There is also the tool box
visible on the left fender.

Painted by Mariusz Motyka

Jagdpanzer 38(*t*) *Hetzer* of one of the ROA (Russian Liberation Army) detachments. Prague, May 1945. The vehicle's camouflage scheme represents the *Škoda* factory-applied variant. There are also Russian national markings painted on the superstructure sides including the white-blue-read tsarist colour combination.

Painted by Stefan Dramiński

Jagdpanzer 38(t) Hetzer captured by the Polish Army during the Warsaw Uprising and named Chwat (*Merry Blade*). Warsaw, August 1944. The vehicle had been captured on August 2nd, 1944 in the Napoleon Square area by the soldiers of *Kolegium C Kedyw* company and 1st and 2nd company of *Kiliński* Battalion. Firstly it was a part of the barricade. Few days later it was repaired but did not take part in the fighting.

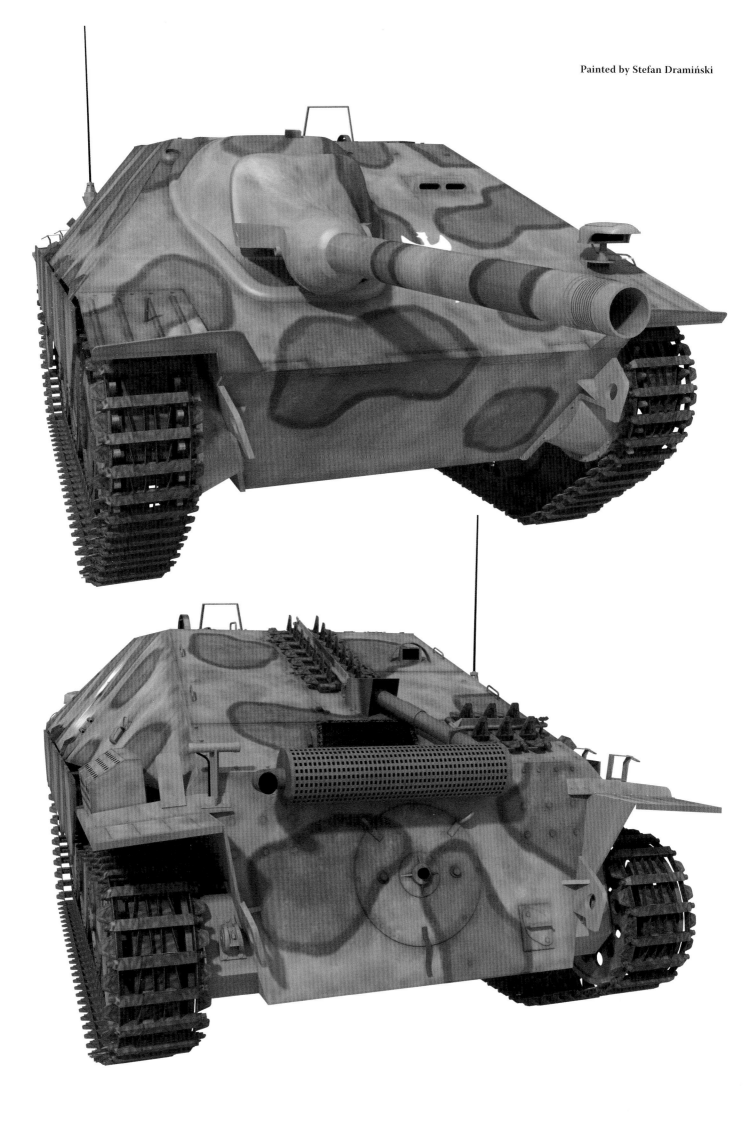

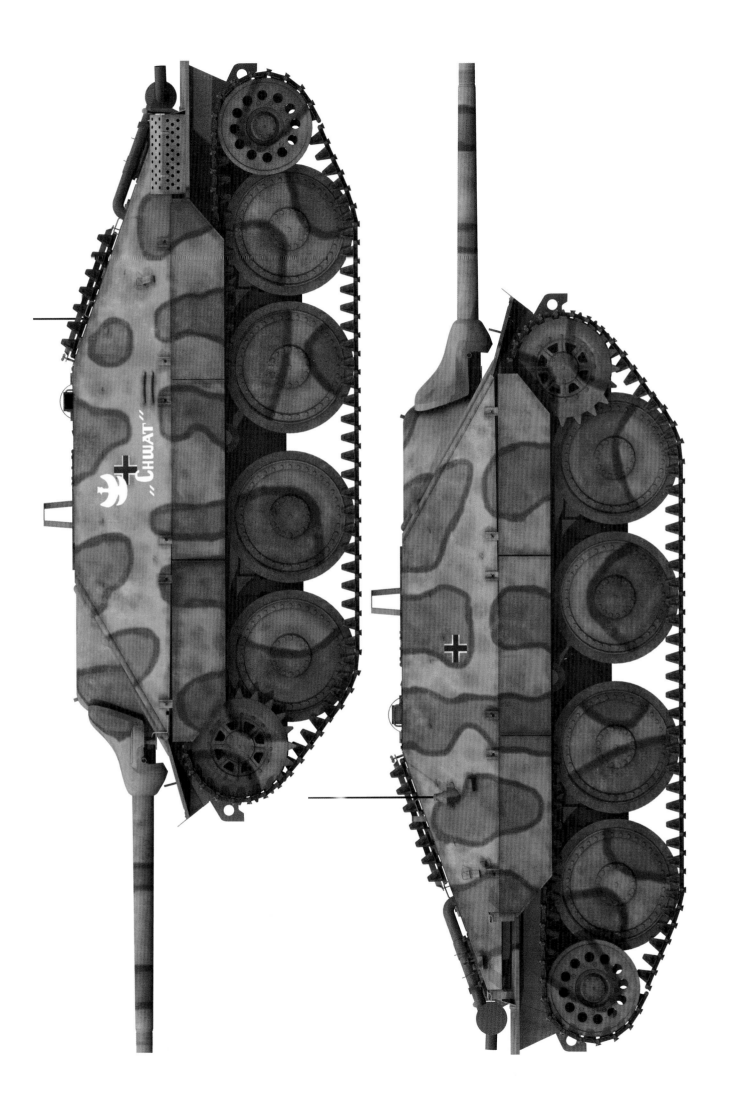

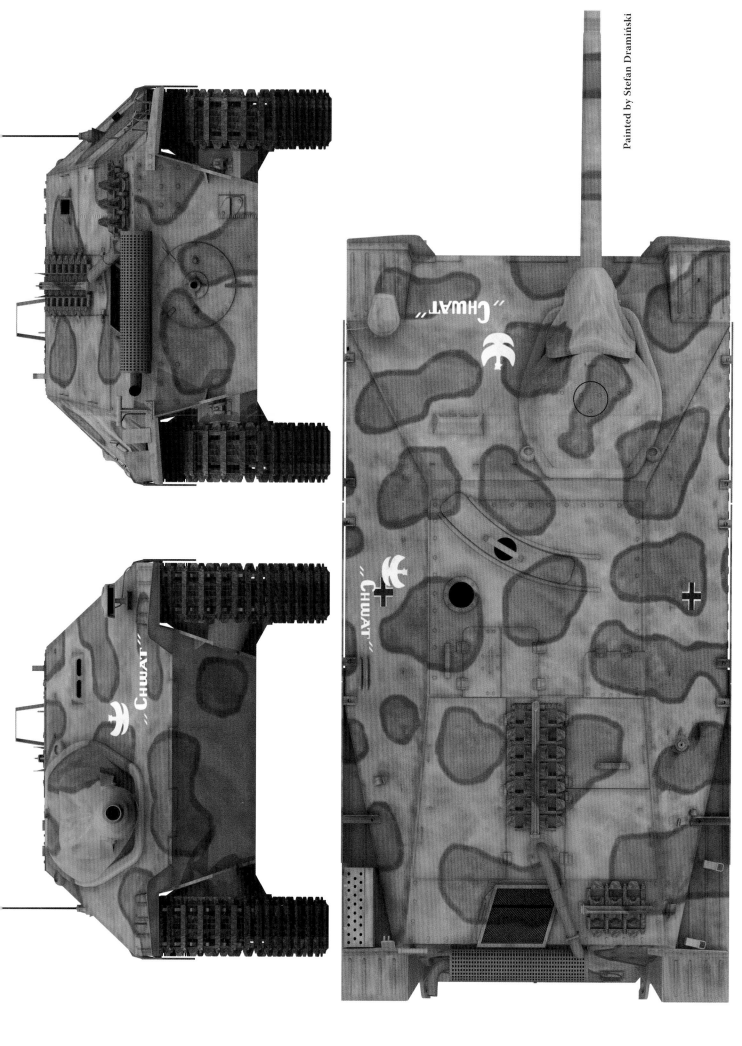

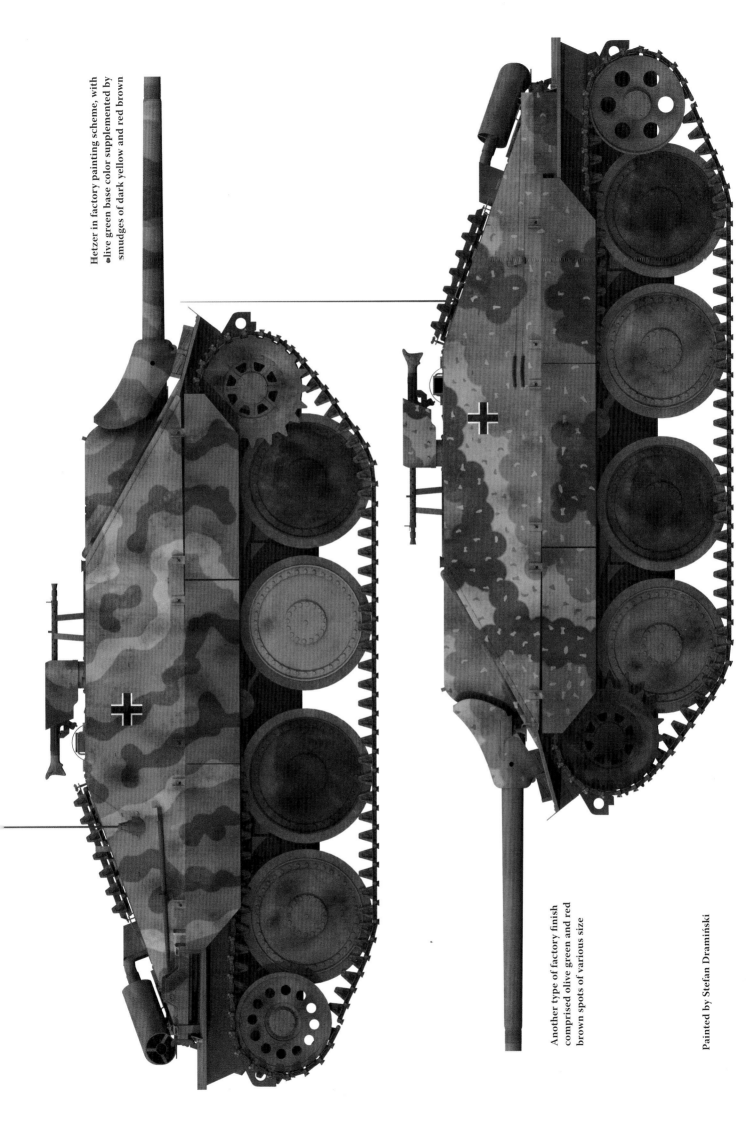

Hetzer in factory painting scheme, with olive green base color supplemented by smudges of dark yellow and red brown

Another type of factory finish comprised olive green and red brown spots of various size

Painted by Stefan Dramiński

Painted by Stefan Dramiński

Hetzer coded "211" of an
unknown unit, Bohemia 1945

Hetzer coded "153"
(previously "003"),
Prague, May 1945

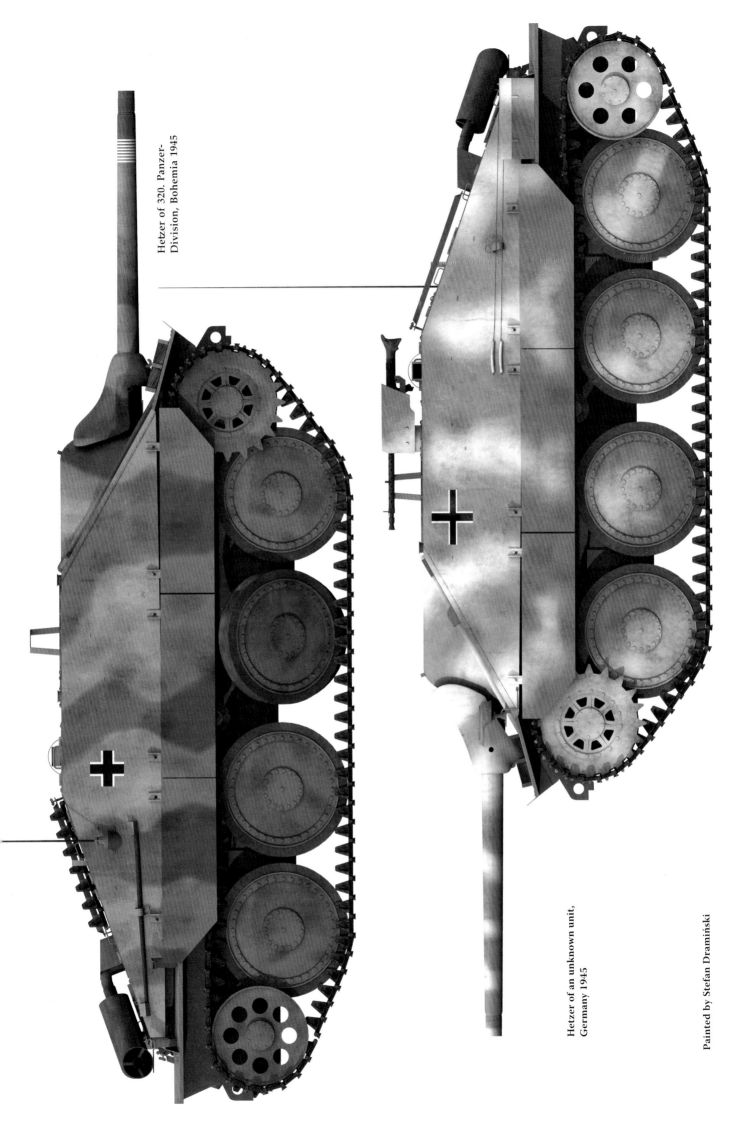

Hetzer of 320. Panzer-Division, Bohemia 1945

Hetzer of an unknown unit, Germany 1945

Painted by Stefan Dramiński

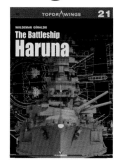

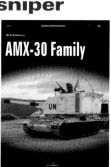
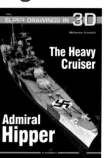